BLOCKCHAIN BASICS

TEEN EDITION

Jacqueline Hayes

Blockchain Basics: Teen Edition/Jacqueline Hayes/ First Edition

Table of Contents

BLOCKCHAIN, BLOCKCHAIN, BLOCKCHAIN

WHAT IN THE WORLD IS BLOCKCHAIN???

I DON'T KNOW!!!!!!!!!!!

INTRODUCTION

Blockchain is one of the newest technologies to hit the world stage.

Blockchain technology will revolutionize how we do business in this country and all around the world.

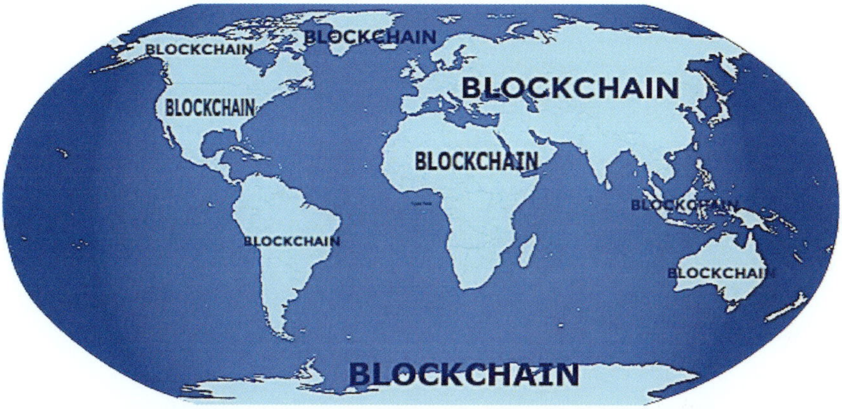

Blockchain is considered to be Internet 3.0. Industry leaders in finance, food supply, energy, healthcare and other areas including government believe that it will change our world as much or more than the internet has over the last two decades. Sometimes you may hear Bitcoin and blockchain used interchangeably, but don't be confused. Blockchain is not a Bitcoin.

Blockchain is the technology used to support Bitcoin transactions. Blockchain itself is a digital database designed to process transactions and store information securely. Blockchain technology can make it faster, cheaper, more efficient, and more secure to transact almost any kind of business anywhere in the world.

It is boasted that blockchain will be a three-trillion-dollar industry. IBM's General Manager of blockchain, Marie Wieck, sites research from Gartner, Inc. which estimates blockchain will add $3.1 trillion to the global economy by 2030. Yes, three trillion dollars! All those zeros are really exciting!

BENEFITS OF BLOCKCHAIN TECHNOLOGY (BCT)

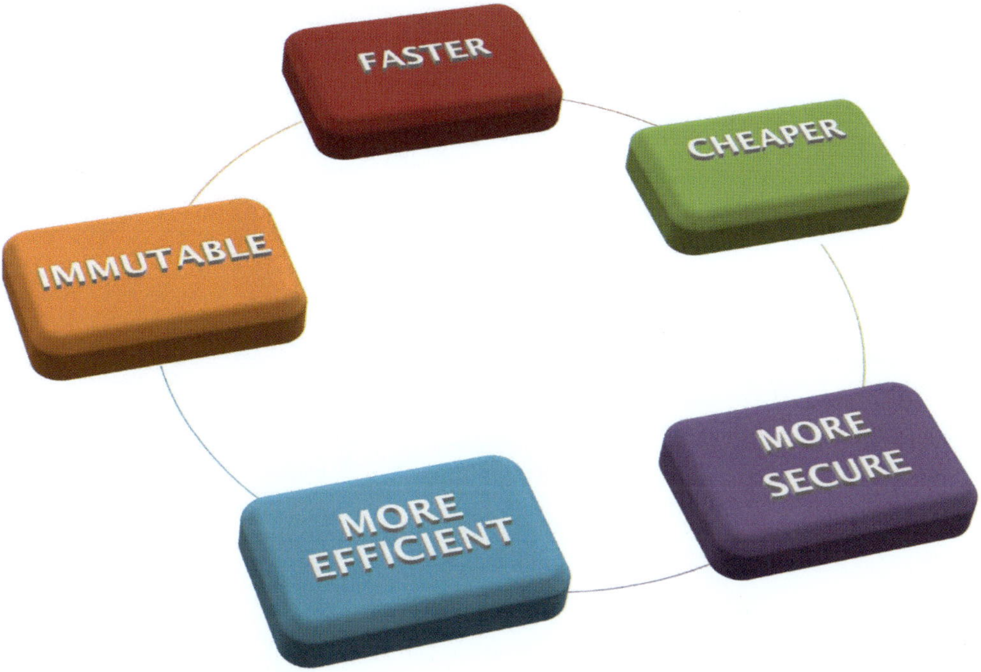

We all live in a global society. We communicate and conduct business all over the world every single day. If we are to keep pace with the growing global economy, we must develop and use technologies that keep us in the mainstream. Blockchain technology does just that thanks to the five reasons above. Let's discuss them further.

FASTER

Blockchain makes the process for completing transactions more streamlined and automated. As a result, data is processed within seconds so results for any transaction is almost immediate instead of taking hours or days. It can speed up processing financial transactions. Also, there is no need to wait for a third party to say your transaction is approved and charge a fee.

CHEAPER

All businesses want to reduce their costs and consumers want to pay fewer fees or no fees at all. Since blockchain data processing platforms eliminate the need for a middle man, it costs less to do business. Blockchain also cuts costs by eliminating a lot of paperwork and computers servers used to process and store information.

MORE SECURE

Blockchain technology offers some very trustworthy security measures:

1. Cryptography is used to encrypt or code all information on the blockchain, meaning it turns plaintext into computer code.
2. No one can randomly change, delete, or misplace information on the blockchain. If a change is needed, a whole new block has to be created and its contents have to be approved by the network participants. The block with the incorrect information remains part of the blockchain.
3. Since blockchain does not have one central authority running everything, information is not stored in just one location. It is stored in digital ledgers and each node on the blockchain has a copy of the ledger.
4. Can blockchain be hacked? A hacker would have to take over 51 percent of the nodes in the network. There can be hundreds of nodes and they can be anywhere since there is no centralized system. It would be very, very difficult for hackers to invade a system and steal or compromise the information.

CRYPTOGRAPY
ALGORITHM
ENCRYPTION
DIGITAL
SECURE
ATTACK
HASH
PUBLIC
CRYPTO KEYS
PLAINTEXT
CIPHERTEXT

IMMUTABLE

Once a block of information is created and added to the blockchain, the information in that block cannot be changed. The information (data) is immutable (unchangeable).

If an error is discovered, a whole new block must be created containing the correct information. Just to make sure everyone knows this new block is correcting an error in a previous block, the new block has to reference the block that contains the error. Both blocks remain part of the blockchain forever.

MORE EFFICIENT

When businesses use the old fashioned, paper-pushing method to get a transaction completed it's a long process, costs money, and there is always a chance for human error.

Blockchain uses a digital super highway to process transactions quickly. The individuals creating the blockchain network simultaneously work to verify transactions, so you don't have to wait hours or days for a transaction to be completed. Intermediaries—the famous middle man—are not needed, which results in transactions being completed at less cost and much faster.

Since the process is done by a computerized mathematical system, human error is extremely rare.

Blockchain also eliminates the need to trust those mysterious behind-the-scene people who work on your transaction request. Who are they? Why should you trust them to handle your personal information? No need for that with blockchain. Requests processed on the blockchain are encrypted and securely stored on the blockchain. Hackers don't have a chance.

Because of these features, blockchain can be used by consumers, businesses and government agencies to:

- eliminate fraud on luxury items (Prada, Gucci, Burberry, Hermes), fine art, and pharmaceuticals
- protect your identity
- make international payments or trade stocks, bonds, and commodities
- allow musicians, writers, and other artists to pay themselves every time their work is bought or resold
- vote in elections

It is very difficult to hack a blockchain. As mentioned before, a hacker would have to get 51 percent of the people working on the network to agree with his action. That could be hundreds of people. Then every block on the chain would need to be changed to match whatever the hacker had done. It's quite a difficult task.

The result with blockchain is your transactions get done faster, cheaper, securely, and with very little chance for hackers to mess with your information.

HISTORY LESSON

Before we get into all the technicalities of blockchain, let's have a short history lesson and look at the road that led to the blockchain technology of today.

The internet has been around since the 1950s and was used to share information between computers, primarily by the government, scientists, and researchers. In 1989, the World Wide Web was invented by Tim Berners-Lee for public use.

Actual commercial use of the World Wide Web started in the mid-nineties. It changed how we make payments, our mobile service, and the entire marketplace. As the internet continued to evolve, it opened the door for more and more innovative uses.

This table shows the roadmap:

1990 - 1995

- first website and server
- first internet search tool
- web made available to the public
- the White House comes online
- creation of Amazon, Craigslist, Match.com, eBay, Windows 95, first Internet Explorer, Yahoo!

1996 - 2000

- rise of Internet 2.0 and web-based commerce
- creation of PayPal, Netflix, GoDaddy, Google, Napster
- first blog
- news media online
- 41 percent of adults use the internet

2001 - 2005

- creation of Wikipedia, Facebook, and LinkedIn
- Political campaigns
- Amazon Prime
- Etsy
- Xbox Live

2001 - 2005

- iTunes Music Store, Skype, Myspace, WordPress
- 55 million people use the internet at work
- Firefox 1.0
- YouTube, Reddit

2006 - 2011

- Twitter
- First iPhone/Chrome web browser
- Apple app store
- **2008 INVENTION OF BLOCKCHAIN**
- Bing search engine
- Viral videos
- Pinterest, Instagram, Quora, Google+
- #use of hashtag

In 2008, few had confidence in the viability of blockchain. Since then, blockchain has managed to capture the attention of major industries, giant corporations (IBM, Amazon and Microsoft), and local and federal governments. That's just in the United States. To name a few international places Nigeria, Dubai, India, China, Switzerland and the United Kingdom are also accepting and using blockchain as well. The global race is on to be the first blockchain city, the first blockchain state, and other first-time blockchain innovations. Dubai has set 2020 as a target to have the first blockchain powered government.

Since its inception, blockchain has gone through some major upgrades.

- First was the development of the blockchain platform.

- Next came smart contracts which are programs of instruction built into the blockchain. Smart contracts run on the Ethereum blockchain. They are designed to go into action when certain situations occur in a blockchain transaction. For example, let's say blockchain is being used to process the sale of a privately-owned vehicles. The blockchain has a smart contract designed to alert a potential buyer if a vehicle has certain deficiencies and discontinue processing the sale. The deficiencies are clearly defined in the smart contract. In this case, there is a recall on the vehicle for faulty driver side airbags. Because of the smart contract the sale is not processed. The potential buyer escapes a bad deal.

Traditional contracts are written documents maintained in a file cabinet or a data base. Someone has to prepare them, maintain them and enforce them. In many cases the terms of the contract have to be interpreted for all parties involved. Each of these actions requires a person and a fee. This process is time consuming and costly.

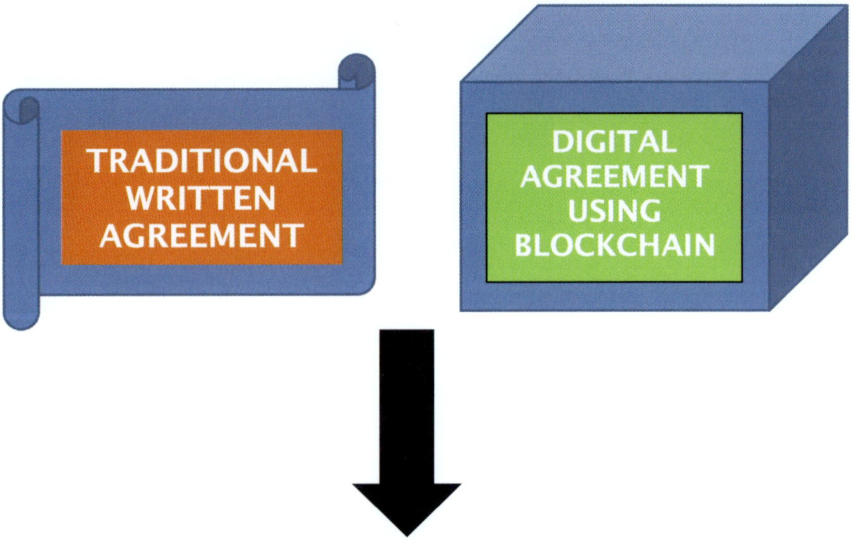

- Currently, the major hurdle for blockchain is scaling which is the ability of the blockchain to handle an increased workload. Can blockchain keep up with a growing and expanding business? Will blockchain be able to process more transactions at the same or faster speed and still maintain the highest level of security? Those are the questions being addressed now. If you are a business owner, one of your major concerns is the ability of your technology to keep up as your business expands. You don't want your business to stall because your technology can't maintain the pace.

Since its inception, numerous blockchain platforms have been developed. Some still exist and some have faded away. A few of the major platforms are: Ethereum, Hyperledger, Corda, EOS, Dragonchain and VeChain.

Before we begin to explore other aspects of this revolutionary technology, let's look at who invented blockchain. In 2008, a person or group of people only known by the pseudonym, Satoshi Nakamoto, invented blockchain.

He or they invented blockchain to support Bitcoin (cryptocurrency or digital currency) transactions. Blockchain also addressed the problem of spending a digital currency twice. Since Bitcoin is digital currency it can be reproduced. Remember blockchain and Bitcoin are not the same thing, Blockchain is the operating platform for Bitcoin. "Blockchain is like the highway and Bitcoin is like the car." (Cheree Warrick, Founder of Black Blockchain Consultants).

Just like Apple iPhone, iPad and iPod use the IOS platform to operate, Bitcoin uses blockchain. Today, the use of blockchain reaches far beyond Bitcoin and other cryptocurrencies.

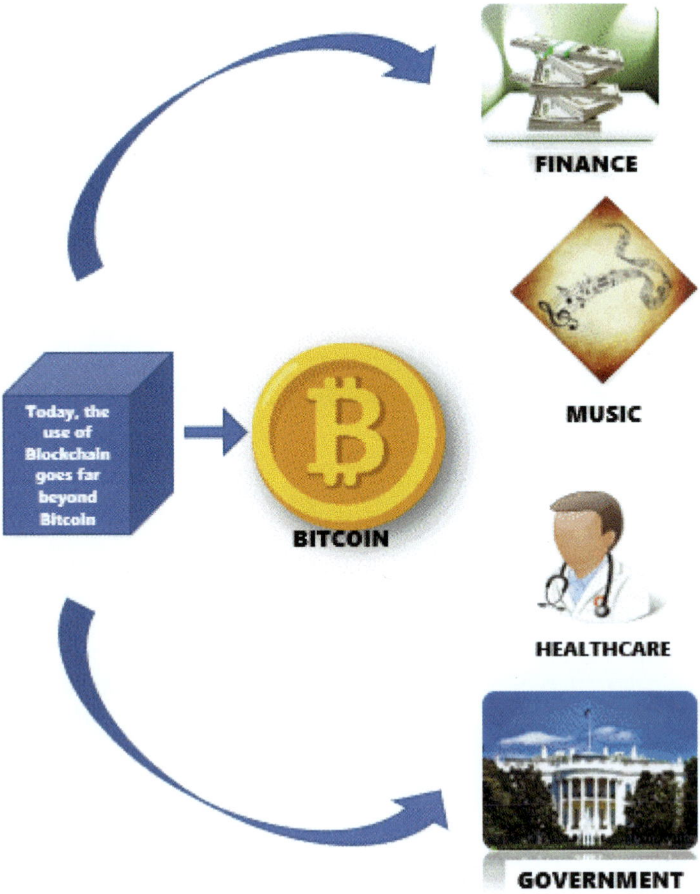

Blockchain can be used in all the areas above, as well as supply chain management, mobile payments, money transfers, titles, securities, identity verification, voting, music copyright registration, update student accomplishments, car rentals and cryptocurrency trading. The list is endless.

BLOCKCHAIN FACTS

Let's review a few facts about blockchain.

- Blockchain is a new and evolving technology
- Blockchain was invented in 2008 by Satoshi Nakamoto
- Blockchain is boasted to be Internet 3.0
- The internet and the World Wide Web were invented before blockchain
- Blockchain is not a cryptocurrency
- Blockchain is faster, cheaper, and more secure than the current technology used in business and consumer transactions
- Blockchain is immutable
- Blockchain uses smart contracts
- Blockchain has more uses than originally intended
- What do you remember about how blockchain can be used?

____

____

____

WHAT IS BLOCKCHAIN TECHNOLOGY?

Blockchain is very simply a digital record called ledgers. It is also referred to as Distributed Ledger Technology (DLT). Blockchain can securely process, record, and store transaction information or track any asset whether tangible, intangible, or digital.

Here are two ways to think about blockchain in real terms vs the digital space:

It's like a great big file cabinet.

- Each file drawer represents a block on the blockchain.
- Once the drawer is filled with files, it is closed and locked and becomes a part of the entire file cabinet.
- So, when a block is full of verified transactions, the block is securely added to the blockchain.

OR

Think of blockchain like a freight train.

- The blocks on a blockchain are like the railcars on a train.
- Each railcar of a freight train is labeled with identifying information.
- Each block on the blockchain has identifying information.
- Once the rail car is loaded with various kinds of cargo, the railcar is closed, locked, and added to the rest of the train.
- Once a block is loaded with encrypted data and validated, the block is added to the blockchain.

As you learn more about blockchain, think of a way you can explain blockchain to someone and help them understand what it is. Compare it to something you are familiar with.

The actual blockchain is a countless number of data blocks (digital record) that are added to the blockchain in chronological order. The very first block in a blockchain is called the genesis block.

Not every blockchain platform works the same because blockchains are designed to facilitate a particular business operation.

The unique thing about blockchain technology is the way information is processed, validated, and stored. It uses a network of computer systems to perform these tasks. The network is made up of individual computer systems called nodes.

There are three types of networks:

- Centralized
- Decentralized
- Distributed

CENTRALIZED NETWORK

Many businesses and government agencies use a centralized network to process and store data. In a centralized network all the computers, or nodes, are in a fixed location, and they are controlled by one central server. The central server is the root node. The other nodes are supervised by the central server. They are assigned tasks and cannot work independently on any transaction or communicate directly with other nodes. Also, all the data is stored and maintained at a central point. If the central server breaks down, all the other nodes in the network also become inoperable. Most banks use a centralized system.

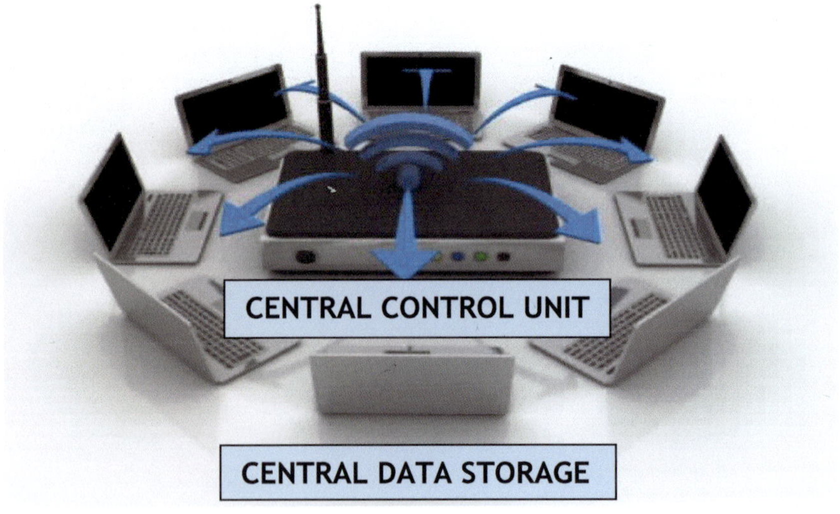

DECENTRALIZED NETWORK

A decentralized network is very different from a centralized network. It has several root nodes. Decentralized networks do not rely on one central point of control. Each node operates independently and can communicate with the other nodes in the network. If a node shuts down, it does not affect the other nodes in the network. All data is stored on the blockchain rather than one system like the centralized network. Bitcoin operates on a decentralized blockchain network.

DISTRIBUTED NETWORK

A distributed network is considered to be a peer to peer (P2P) network because the node operators have equal capabilities and responsibility for processing transactions. They share information, resources, access to transactions and copy and share data across the network. The nodes are generally spread out geographically. Data is stored on the blockchain. Airlines use distributed networks.

In short, centralized and decentralized refers to who has control of the networks. A distributed network is about the location of the network and it can be centralized or decentralized.

Share your thoughts about what you have learned about blockchain technology so far:

HOW DOES BLOCKCHAIN WORK?

All blockchains do not work the same way. Each blockchain is designed to meet a specific need or solve a specific problem.

There are some technical terms in this section on blockchain. You may or may not remember all of them, but it is important to remember the flow of the process. Below are some terms we will use in this section to refer to as needed.

- Algorithm: a computerized mathematical process used to solve a data-processing problem.
- Hash: a unique digital fingerprint that identifies a block and its contents
- Hashing: a process that uses an algorithm to change the information that is entered into the system (input data).

Once the information is processed through the algorithm, the output is a unique string of numbers and letters (computer code). That output data or computer code is known as a hash. No matter how large the input data, the result or output data is always a 32-byte hash.

Let's say you initiate a transaction to send $50 to a friend. Your request to "send $50 to Mike" is changed from plaintext to a computer code.

Any change in your input, no matter how small, produces a different output.

PLAINTEXT INPUT HASH TEXT OUTPUT

- Target Hash: guideline for miners also known as node operators. A miner has to get a specific hash to add a block to the blockchain. That hash must be less than or equal to the target hash.
- Mining: node operators perform a task called mining to process transactions and add a block to the blockchain.

If a miner gets a hash that is larger than the target hash, he/she must rehash the original data all over again. The target hash is changed periodically by the network.

> TARGET HASH: 00000031287mbr6549041pj2764wz09sl76548899
> MINER'S HASH: 00000031387mbr6549041pj2764wz09sl76548899 = incorrect hash

Remember all hashes are 32 bytes

- Nonce: any random number selected by the miner to help him/her get the target hash. The Nonce starts at zero and continues up numerically. The nonce can only be used once. (The word NONCE even has the word ONCE in it.)

Since miners use a trial and error method to get the target hash, they often have to repeat the hashing process multiple times. Each time a miner has to repeat the hashing process, he/she adds a random number (nonce) to his/her original input. Adding the random number insures a miner will get a different result, but he/she cannot use the same number more than once.

- In the example below, the target hash is:

  000000312876549041276409765488999

- Miner's first attempt gets an incorrect hash:

  0000629n387654904127640976548899

To rehash the data, the miner adds a nonce, or "0", to the data he/she originally input. The addition of the number "0" results in a different hash. This time he/she gets it right.

- 000000311876549041276409765488898 (This hash is less than the target hash, so it is acceptable.)

What Is a MerkleTree and a Merkle Root?

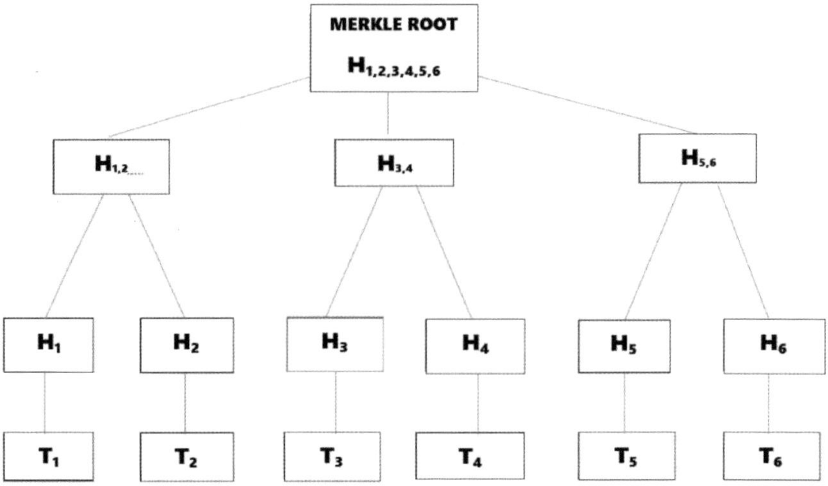

Each block on the blockchain contains a Merkle tree which represents all the transactions in that block. A Merkle tree is read from the bottom up. In the example above, each "T" at the bottom represents one transaction. Each "H" represents a hash of the data below it.

- T_1 thru T_6 represents a transaction on a block
- H_1 thru H_6 is a hash of T_1 thru T_6
- $H_{1,2}$ is a hash of H_1 and H_2
- $H_{3,4}$ is a hash of H_3 and H_4
- $H_{5,6}$ is a hash of H_5 and H_6
- $H_{1,2,3,4,5,6}$ is a hash of $H_{1,2}$, $H_{3,4}$ and $H_{5,6}$
- $H_{1,2,3,4,5,6}$ is a hash of all the transactions in this block. It is called the Merkle root. The Merkle root is the data put on the block instead of that whole list of transactions.

Remember all data on the blockchain is encrypted

This is the process:

Person A initiates a transaction on the blockchain. It could be a request to send money, buy or sell Bitcoin, register a copy of your song, register land, update a student's record, vote, or learn where the beef on your burger comes from.

- The message is encrypted (turned into code) and changed from plaintext to computer code.

- The request goes into a transaction pool with other requests.

- A miner selects the transactions that he/she wants to work on. More than one miner can select the same transaction. If a transaction is not selected it stays in the pool.

Not this kind of miner. . .

. . . this kind of miner

Miners working at network nodes

- Miners participate in solo mining or pool mining. Solo miners work alone and collect the reward. Pool miners work together and split the reward. Miners use various types of equipment for the mining process. They need computing power that is fast and efficient. Some use CPU (slow, little power, inefficient), GPU (slow, uses graphic cards), FPGA (faster than CPU & GPU), or ASIC (fastest).

- Software is used to connect the node operators to the blockchain and the transaction pool. The software enables several important tasks:
    - monitors the mining process
    - displays vital system statistics
    - delivers work to the node operators
    - facilitates the hash process

- Now the race is on to solve the puzzle (create the correct hash). The first node operator to solve the puzzle gets the reward.

- Information about the transaction goes into a temporary block called a candidate block.

- The candidate block is labeled in computer code with the following information which makes up the block header.
    - Version (software version)
    - Hash from the last block
    - Transactions
    - Time
    - Target Hash

- The node operators begin mining. Mining is also known as Proof of Work (PoW), Proof of Stake (PoS), Delegated Proof of Stake (DPoS) or Proof of Weight. When node operators input data into the computer, the computer scrambles the data through a mathematical algorithm, and it produces a hash which must be at or below the target hash.

- The hash is an identifier similar to a signature or fingerprint. The hash is compared to the target hash. If this hash is at or below the target—success! If this hash is above the target, the miner must keep trying. The first node operator to get the target hash is the winner and gets the reward.

- To try again, the node operator must add a random number (nonce) to the original input data—the block header information. The node operator can pick any number starting with zero, but it can only be used once. This random number is added to the block header info and rehashed. This small variation in the input will result in a different hash. The node operator hopes this will result in the target hash.

- The first node operator who guesses the correct nonce that results in the correct hash gets rewarded. The reward is given in cryptocurrency, like Bitcoin for example.

- When the target hash is achieved, the node operator announces his/her results to the other operators on the network. They verify his/her input and output using the algorithm process. It is necessary for the other node operators to validate the work before a block can be completed and added to the blockchain.

- Once the other node operators verify the work and agree that the results are valid(consensus), the block is updated with additional information and added to the end of the blockchain. Cryptography (computer language code) links the blocks together. The updated information includes the nonce and Merkle root. Each node has a copy of the blockchain record (ledger).

- This seems like a long process, but it all happens in about ten minutes.

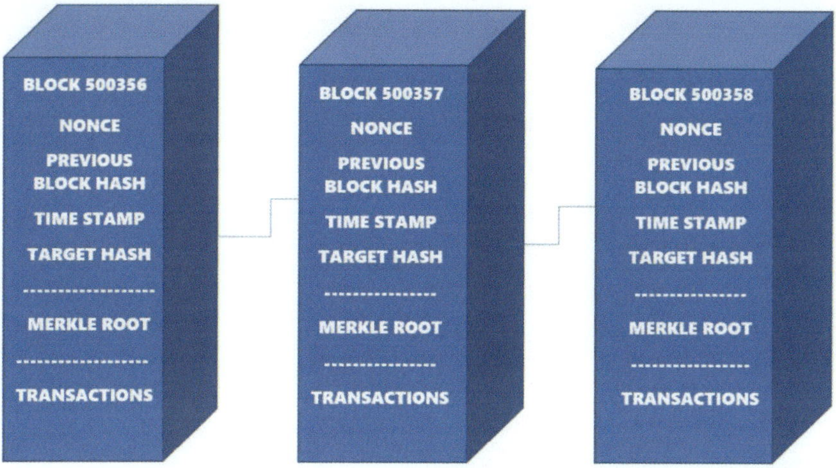

*Note: the block number, nonce, previous block hash, time stamp, target hash, Merkle root, and transactions are in a hash format (32 bits entry) on the actual blockchain blocks. Plaintext labels are used here so you know what each of those hashes would represent.

Blockchain Flowchart:

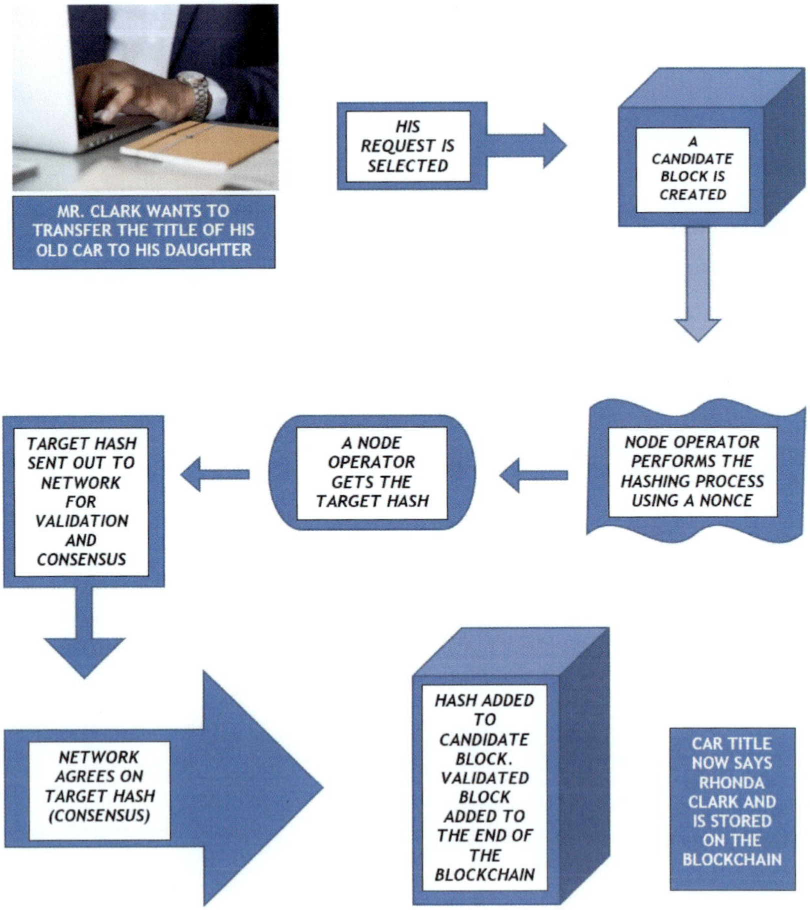

TYPES OF BLOCKCHAINS

There are three kinds of blockchains:

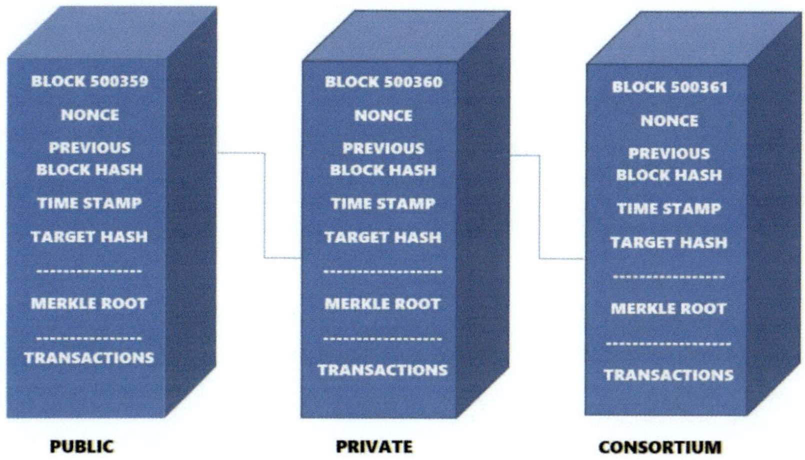

Public: No one owns a public blockchain and anyone on the network can access the blockchain to read, write, review, audit, and add a block to the blockchain. There are no access restrictions. Transactions are visible to anyone, but the user remains anonymous. Bitcoin runs on a public blockchain.

Private: Private blockchains have access restrictions. Whomever owns a private network owns the majority of the nodes. A network administrator must invite an individual to join the network and decides what actions the nodes can perform. Participants on the network can write data but access to read data may be public or restricted.

Consortium: Consortium blockchains have restricted access. The network is managed by a group of companies or their representatives. These individuals make all decisions about the network.

The main difference between these three types of blockchains is (1) who is allowed to participate on the network, (2) who maintains the ledger of transactions, and (3) how they arrive at consensus.

The similarity between these three types of blockchains is that all three are immutable.

CHARACTERISTICS OF BLOCKCHAIN

Permissioned

Some blockchains require consent or authorization for nodes to operate on a blockchain. These are known as permissioned blockchains. They are managed by an individual or group of companies who gives permission to certain nodes for access to the network. They also decide what tasks the node operators can perform. Information is shared with the network on a need-to-know basis.

Permissionless

Some blockchains do not need anyone's consent or authorization for access. Permissionless blockchains are decentralized and not controlled by any single individual or group. Any network participant can access the blockchain and perform any of the blockchain functions.

	PUBLIC	PRIVATE	CONSORTIUM
NETWORK	Decentralized	Decentralized	Centralized Distributed
ACCESS	Permissionless	Permissioned and requires invite from admin	Permissioned with restrictions
TASKS	Read, write, audit, add blocks	Decided by admin	Decided by owners or their reps

Digital Keys

Digital keys allow a customer to access the blockchain network to initiate and complete a transaction. The keys are cryptographically generated in pairs as a public key and a private key. Each key has a digital address. The key address is made up of a long list of numbers and letters (i.e., pajr9nsimfuwm4enMvcqi9Lsaiebg4qpzmalwoKs). These alphanumeric characters are known as the key address. The private key address is the longest.

The public key allows a user to start a transaction on the blockchain and more than one individual can use it. The private key allows a user to complete a transaction. Private key information should never be shared.

WHY?

Let's use a Bitcoin example: Person A is sending Person B some Bitcoin. Person A uses their public key to start the transaction. When the transaction is processed, Person B uses their private key to collect the Bitcoin. Private key information should never be shared. If someone else has the private key information, they could use the key to steal your Bitcoin.

Describe what you understand about blockchain technology so far:

BLOCKCHAIN LANGUAGE

How does blockchain communicate? How do node operators communicate with the computer system? Does the computer system speak English or some other language? Well, it's a little of both. The blockchain network takes our plain text and turns it into computer language. The blockchain network uses different computer languages to operate in the digital space. There are several computer languages in use now:

- C++ is one of the most used languages for programming and has been around a long time.
- Simplicity lets blockchain developers use code for smart contracts
- JavaScript is very common and is used on the internet and by blockchain developers.
- Python is similar to JavaScript and is used by data scientists and analysts.
- Solidity is a very high-level computer language and is used on the Ethereum blockchain.

LET'S PUT IT ALL TOGETHER

Since blockchain was originally invented to support Bitcoin, on the next page, we'll look at a Bitcoin transaction.

You can search for Bitcoin transactions and blocks at https://www.blockchain.com/explorer

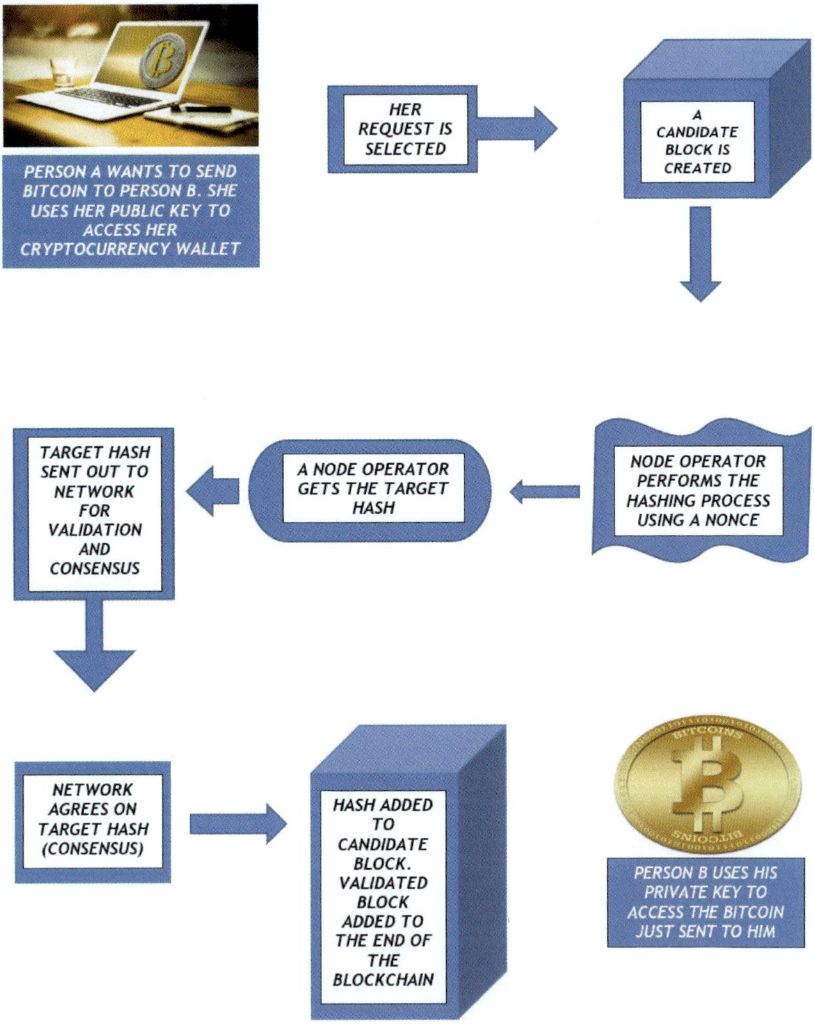
PERSON A WANTS TO SEND BITCOIN TO PERSON B. SHE USES HER PUBLIC KEY TO ACCESS HER CRYPTOCURRENCY WALLET
HER REQUEST IS SELECTED
A CANDIDATE BLOCK IS CREATED
NODE OPERATOR PERFORMS THE HASHING PROCESS USING A NONCE
A NODE OPERATOR GETS THE TARGET HASH
TARGET HASH SENT OUT TO NETWORK FOR VALIDATION AND CONSENSUS
NETWORK AGREES ON TARGET HASH (CONSENSUS)
HASH ADDED TO CANDIDATE BLOCK. VALIDATED BLOCK ADDED TO THE END OF THE BLOCKCHAIN
BITCOINS
PERSON B USES HIS PRIVATE KEY TO ACCESS THE BITCOIN JUST SENT TO HIM

WHERE DO I FIT IN?

Careers and Business Ventures

Blockchain is a new and emerging technology. It has the potential to transform businesses and impact every industry. The need for skilled and knowledgeable blockchain experts will greatly increase as established businesses and entrepreneurs adopt blockchain technology. Blockchain jobs will become more common in the near future.

Almost 50 percent of tech companies are expected to implement blockchain in the next three years. Some industries are already spending lots of money on blockchain technology projects. Financial services, manufacturing, and human resources are leading the way. It is predicted that nearly three billion dollars will be spent this year on blockchain technology. Blockchain will make it possible for businesses to offer new services that were not previously possible.

Various specialists will be needed to:

- study a business and recommend an appropriate blockchain platform
- design a blockchain platform that addresses current and future business needs
- update the blockchain to accommodate a growing and possibly changing business
- interpret and analyze complex digital data

- mine blockchain transactions
- teach blockchain technology
- monitor, test, and inspect blockchain operations
- monitor cyber security
- market a blockchain business
- assist state, federal, and international governments in developing blockchain platforms to solve problems and increase efficiency

Of course, this is a limited list. More and more careers and business ventures will be created as blockchain becomes more of the norm in everyday operations. There will be numerous opportunities for entry-level, intermediate, and management positions and careers. It is not necessary to be a computer coder or have a college degree to secure a blockchain position.

Blockchain as a Service (BaaS) is an interesting and growing field. This service is expected to grow to around thirty billion globally by 2024. BaaS Specialists provide a unique service to their clients who want to use blockchain technology. Instead of building a whole new blockchain, these specialists already have the infrastructure and initial configuration in place. The existing blockchain is only tweaked to fit the client's exact needs, which saves the client time and money.

Blockchain is expected to be a three-trillion-dollar industry. These are some of the career opportunities that are available right now:

- Blockchain Developers/Architect: responsible for the research, analysis, design, and execution of the blockchain
- Blockchain Network Operators (Miners): individuals who process, validate and record transactions to blocks on the blockchain
- Blockchain Intern: a trainee who works on various blockchain projects. Having a background in computer science, engineering, or information technology is very helpful in securing an intern or entry-level position.
- Blockchain Project Manager: an individual who oversees the planning and execution of blockchain projects
- Blockchain Quality Engineer: monitors the quality of blockchain operations
- Blockchain Legal Consultant and Attorney: an individual who specializes in the legal aspects of blockchain technology
- Blockchain Engineer: builds blockchain applications and protocols
- Blockchain Consultant: provides advice about blockchain solutions for business operations
- Blockchain Information Security Analyst: an individual who analyzes security and designs, then implements security solutions
- Blockchain Data Scientist: an individual who authenticates, tracks and reports blockchain data

- Blockchain Market Strategist: builds strategies for attracting and retaining customers
- Blockpreneur (Blockchain Entrepreneur): a person who designs, launches, and runs a business using a blockchain platform
- Blockchain Generalist: an individual who is competent in several areas of blockchain technology (e.g. Smart Contract Specialist)
- Cyber Security Specialist: an individual who ensures the safety of the computer networks and systems

Benefit from the Blockchain Job Boom

According to the Entrepreneur online magazine, "blockchain related jobs are the second fastest growing in today's labor market."

Since the demand is so high, companies are always looking for both technical and non-technical professionals. Some companies are even willing to hire interns and train them on the job.

The Blockchain Training Alliance has an illustration that shows how to plot a blockchain career path for both technical and non-technical blockchain positions. This illustration can be viewed at the link below:

https://blockchaintrainingalliance.com/blogs/news/blockchain-career-paths

Following is a very general guide for starting out in blockchain.

- Understand blockchain fundamentals
- Use free resources to learn the basics (YouTube videos)
- Become part of a blockchain community
- Do freelance work to build a portfolio
- Volunteer with blockchain associations
- Start a blockchain club at your school or with friends

As you gain knowledge and skills, more advanced training may be necessary. At some point you may be interested in a specific area and decide to become a specialist in that area.

Education and Training

Blockchain is catching on around the world. Many institutions have begun to realize that there is a growing need for education and training on all aspects of blockchain technology. Some universities around the world have begun to offer at least one course in blockchain technology and the majority of these universities are in the United States. There are also online training programs and online tutorials available. Blockchain Consortium International offers several courses through their Center for Blockchain Studies at

www.blockchainstudies.tech.

It may seem as if there are few options for education and training, but we are in a world that is overflowing with information.

It may be necessary to join blockchain organizations, attend conferences, network with blockchain enthusiasts, look at videos and read as much as you can about this subject. Find the area that interests you most and absorb everything you can.

Institutions that offer some blockchain classes or certification:

- ❖ Cornell University
- ❖ Duke University
- ❖ Georgetown University
- ❖ New York University
- ❖ Princeton University
- ❖ Stanford University
- ❖ University of Nicosia – Cypress
- ❖ University of Illinois – Urbana Champagne
- ❖ University of California – Berkeley
- ❖ Massachusetts Institute of Technology (MIT)
- ❖ Royal Melbourne Institute of Technology - Australia (RMIT)

Industries currently using blockchain technology:

- Cyber Security
- Finance
- Healthcare
- Marketing
- Education
- Government
- Human Resources
- Real Estate
- Entertainment
- Sports
- Manufacturing
- Entertainment

Top companies using blockchain:

- Amazon
- Daimler Ag
- Hitachi
- IBM
- Nasdaq
- Mercedes Benz
- Microsoft
- Walmart
- Maersk

International places using blockchain:

- Australia
- Bermuda
- China
- Cyprus
- Dubai
- Estonia
- Georgia
- Ghana
- Gibraltar
- India
- Kenya
- Liechtenstein
- Malta
- Nigeria
- Russia
- Rwanda
- Singapore
- Slovenia
- South Africa
- Sweden
- Switzerland
- Uganda
- United Kingdom
- United States

How do you plan to enter the blockchain space?

- You could build blockchain platforms for new and existing businesses.
- You could monitor blockchain operations for a Fortune 500 company.
- You could be a cryptocurrency miner.
- You could analyze blockchain data for small businesses or a large corporation.
- You could design and implement smart contracts for multiple business uses.
- You could be a blockchain consultant and advise various industries and government organizations on the viability of blockchain for their particular situation.

ATTENTION! ATTENTION!

The field is wide-open and opportunities are endless. There are plenty of opportunities in blockchain that do not require any computer coding skills. Find an area that interests you and jump in!

If you are interested in healthcare, study your healthcare area of interest and find out how blockchain can be used in that area. The medical profession can use a technology like blockchain to manage patient information so there is no need for paper records or databases that are controlled and maintained by a centralized network.

WHAT DO YOU REMEMBER?

In your own words describe a blockchain and what it can be used for?

Who invented blockchain? When?

Are blockchain and bitcoin the same thing?

What is a major hurdle for blockchain?

Name three industries that are currently using blockchain?

Describe the three basic networks? What kind of network does blockchain use?

What are three benefits of blockchain and how do they work?

What is an algorithm?

What is hashing?

What is a target hash?

What is a nonce?

What is a Merkle root?

What information is in a block?

Are blockchain miners and node operators the same?

True or false? Node operators are limited to only one attempt to get the target hash.

Briefly describe the process to get the correct hash.

What are the three kinds of blockchains? How does each work?

Is it necessary to keep your digital key information secure? Yes or No Why?

Is blockchain technology used all over the world or just in the United States?

Name two blockchain computer languages.

Where can you get education and training in blockchain?

I'VE GOT IT!!!!!!!!!!!

GLOSSARY

Algorithm: a computerized mathematical process used to solve a data processing problem

Bitcoin: cryptocurrency invented by Satoshi Nakamoto

Blockchain: a digital ledger of transactions

Centralized network: a computer network controlled by a central authority

Consensus: a protocol that requires each node on the network or the controlling entity to agree on the target hash before a block can be added to the blockchain

Consortium: a group with the same business interests

Cryptography: the process of transforming plaintext into an unreadable format (code)

Cryptocurrency/digital wallet: a software program that stores public and private keys, interacts with the blockchain and allows users to send and receive cryptocurrency

Decentralized network: a computer network without a central controlling authority

Decrypt: change coded text into plaintext

Digital: values represented by zero (0) and one (1)

Distributed Ledger Technology (DLT): a digital database that is shared and synchronized across multiple sites

Distributed network a network spread over several computer nodes that share resources

Encrypt: change plaintext to coded text

Genesis block: the very first block in a blockchain

Hash: digital fingerprint that identifies a block and its contents

Hashing: uses an algorithm (computerized mathematical process) to scramble input data and produce a fixed-length output

Immutable: cannot be changed

Ledger: a written or computerized record of completed transactions

Merkle tree: contains all the transactions in a block

Merkle root: one hash that represents all the transaction hashes in a block

Miner: individuals or company who process blockchain blocks

Mining: node operators add transactions to a block which is then added to the blockchain

Node: an operator on a computer network

Nonce: a random number selected by a miner to help him/her get the target hash

Peer to Peer (P2P): two or more computers that share information and resources

Permissioned: a network that requires access authorization

Permissionless: a network that does not require nodes to get authorization to access the network

Private blockchain: requires access authorization to work on the blockchain

Private key: used to decrypt (decode) messages and is always paired with a public key

Delegated Proof of Stake (DPoS): node operators vote to determine who will mine the block

Proof of Stake (PoS): requires miners to have a stake in the process through buying their way in. The stake is the number of crypto coins in the miner's holding account.

Proof of Weight: miners choose based on the number of coins and the number of files he/she holds

Proof of Work (PoW): requires node operators to use the hashing process to mine blocks on the blockchain and receive rewards

Public blockchain: does not require authorization to work on the blockchain

Public key: used to encrypt (code) messages and is always paired with a private key

Target hash: sets a guideline for the miners

Transaction: the transfer or exchange of goods, services, or funds

REFERENCES

https://en.wikipedia.org/wiki/History_of_the_World_Wide_Web

https://medium.com/@robertgreenfieldiv/explaining-proof-of-stake-f1eae6feb26f

http://www.pewinternet.org/2014/03/11/world-wide-web-timeline/

https://www.ibm.com/blogs/blockchain/2018/02/top-five-blockchain-benefits-transforming-your-industry/

https://learnmeabitcoin.com/

https://www.investopedia.com/tech/how-does-blockchain-work/

https://www.investopedia.com/terms/b/blockchain.asp

https://www.investopedia.com/terms/t/target-hash.asp

https://www.investopedia.com/news/public-private-permissioned-blockchains-compared/

https://medium.com/nakamo-to/whats-the-difference-between-decentralized-and-distributed-1b8de5e7f5a4

https://hbr.org/2017/02/a-brief-history-of-blockchain

https://cointelegraph.com/bitcoin-for-beginners/how-blockchain-technology-works-guide-for-beginners#can-you-use-a-blockchain-as-normal-

https://www.studyinternational.com/news/10-best-universities-to-study-blockchain-degrees/

https://www.theladders.com/career-advice/7-best-blockchain-careers-and-jobs-for-the-future https://bitsonblocks.net/2015/09/09/gentle-introduction-blockchain-technology/

https://www.forbes.com/sites/bernardmarr/2018/06/18/the-best-blockchain-jobs-and-careers-available-today/#4d0a54627ca0

https://blockchaintrainingalliance.com/blogs/news/blockchain-career-paths

http://ibm.co/2olpTNL

https://bitcoinexchangeguide.com/blockchain-distributed-ledger-technology/

https://blockchainhub.net/blockchains-and-distributed-ledger-technologies-in-general/

https://www.quora.com/What-is-the-best-programming-language-to-learn-if-you-want-to-work-on-the-blockchain

https://igniteoutsourcing.com/blockchain/blockchain-use-cases-by-industry/

https://www.technoduet.com/a-comprehensive-list-of-blockchain-platforms/

https://cryptomaniaks.com/latest-cryptocurrency-news/blockchain/blockchain-protocols-learn-them-all

https://money.usnews.com/careers/applying-for-a-job/articles/2018-07-25/how-to-benefit-from-the-blockchain-job-boom

https://www.entrepreneur.com/article/314921

Questions? Comments? Contact me at
info@blockchainlearningcenter.com

Printed in Great Britain
by Amazon

75115646R00041